GUERRA'S

LAWS

GUERRA'S

LAWS

by

Guadalupe Guerra

ISBN-13: 978-0-9766983-1-9
ISBN-10: 0-9766983-1-5

Contents

Preface

After serving six years in the United States Navy, I moved back to Texas and landed a job at a local refinery. After years of working there, I was severely injured in an incident that took place classifying me as "Disabled".

For years, I went through some very hard times, losing my livelihood, and trying to deal with my "disabilities".

Because of my disability, I was no longer enjoying reading or watching a good movie and my train of thought does not allow me to sometimes enjoy or concentrate on playing a good game of chess.

Unable to work, I watched a lot of television, especially the Science Channels and anything to do with the history of science and inventions. What else was I supposed to do with my spare time? I thought to myself.

One day I came across a magazine article covering the perpetual motion machine (PPM) and how it was impossible to invent one, much less patent one. This picked my curiosity and did some in-depth research on the subject.

Well, one thing leads to another and before I knew it, I found myself questioning some of the things that I was watching on the history of science and inventions. I started researching the history of the perpetual motion machines and taking notes so that I could review them later.

It's also known as perpetual motion (PM) and perpetual motion devices (PMD). Supposedly, the perpetual motion machines tend to stay in motion forever while at the same time promising a virtually free and unlimited source of energy. Any attempt to invent such a device of any kind would be considered impossible.

Throughout history, physicists have come up with all kinds of reasons to disprove the impossibilities on creating perpetual motion machines based on such things as the law of conservation of energy, first and second laws of thermodynamics.

I do not know how or when it happened, but it all began by simply designing a toy meant only as a hobby that would imitate a perpetual motion device. Maybe it was because no one has ever been successful in creating one. I thought to myself, this little hobby of mine might be exactly what I was looking for in keeping me busy for quite a long time.

I finally perfected an ideal drawing on paper for a small mechanical device, in hopes that it would imitate perpetual motion.

As I was reviewing my notes, I looked down to double check my schematics of the device on paper, when it hit me. It was only a glimpse at first, but then, there it was. I saw the reason as to why building a perpetual motion machine could not violate any laws of physics.

I was able to see with the knowledge and understanding that "I" would be the one responsible for setting my perpetual motion device into motion. By giving it a beginning, I also was giving it an end.

I also realized that perhaps nobody has thought to notice that in order for perpetual motion devices to be in motion, one would have to create or manufacture such a device thus giving it a beginning. WOW! It was a Eureka moment for me.

I am not a scientist, nor a writer. I am just someone who was excited and nervous about putting my groundbreaking scientific discovery down on paper. I was in total awe.

My hobby started as a challenge and a mystery, to be able to invent a toy that would imitate a perpetual motion device. Now, I did not know who to tell about my groundbreaking scientific discovery.

As for the perpetual motion machines, they do not exist in motion, until they are initially put into motion by some type of external or internal "ON" switch. It would also mean that all perpetual motion machines will eventually come to rest.

Therefore, I started telling my family and friends. At first, my family did not know where I was coming from but, they were supportive and my friends seemed to understand somewhat.

This discovery helped me to have a better understanding to further my research in other fields of science that lead me to discover new laws of nature in mathematics and physics.

I remember as a child, during the fall season, playing outside in front of my grandparent's dirt road, I saw a maple leaf landing beside my feet, as a small gust of wind blew by me.

As I was picking it up, my mind began to wonder as to how far beyond the surface of the leaf does one have to go to see what lies beyond its surface. Now, I understand that one may go as far as one needs to go to find the origin.

Scientific research in the fields of mathematics and physics throughout the years lead to discovering new laws of nature in the 21st Century and simplified in this book.

Where, in today's time, does someone turn to when they discover new laws of nature on mathematics and physics?

Chapter 1

Law of Origin

Law of Origin states:

1. "A body set in motion, from a point of origin, will come to rest".

2. "A body cannot be in motion without a point of origin".

3. "A body cannot come to rest without a point of origin".

4. "A body cannot exist without a point of

origin".

I discovered the Law of Origin while researching the history of the perpetual motion machine.

It was then that I realized that the word "machine" was the Achilles heel of perpetual motion.

The word "machine" gives perpetual motion a beginning and with a beginning, it has an end. In conclusion, a body that is set into motion from a point of origin will eventually come to rest.

Chapter 2

Law of Motion

Law of Motion states: The acceleration of a body is parallel and directly proportional to the net amount of energy applied creating momentum to act on the body, is in the direction of the net energy applied, and is inversely proportional to the mass of the body, i.e., $Ep = ma$.

This new Law of Motion of mine is the update to Sir Isaac Newton's Second Law of Motion and my physics formula $Ep = ma$ is the update to the physics formula $F = ma$.

Meaning that energy (applied energy) is the origin by which all things are set in motion. Once energy is applied, it creates what is known as momentum. When this momentum (object in motion) comes in contact with another object or objects, it makes surface contact, also known as a force.

Based on my finding, force (itself) cannot push or pull an object because it does not mean motion of any kind, it has no substance or matter and it is not even a particle. It is my opinion in physics the word, "Force" is just an expression to express an idea, like one would use the word love to express one's feelings, but in this case, it's used as an expression when making surface contact. When two or more objects come in contact with each other, this force, or surface contact, enables for momentum to push or pull on the object or objects making force an axiom or self-evident that requires no proof.

Without energy, there is no momentum. Without momentum, there is no force.

This also means that the Principle of Similitude or the dimensional analysis tables may need updating.

Chapter 3

Law of Measurement

Law of Measurement states: Dichotomizing the tick mark that represents the number Zero (0) into two equal tick marks will make a half negative number Zero (0) tick mark and a half positive number Zero (0) tick mark.

For the purposes of this discussion on my discovery, I will be using the term "tick mark" instead of terms called "hash mark", "hatch mark" or "tick".

While researching the number Zero (o) using Real Numbers on a number line, lead to the discovery of the Law of Measurement.

I notice that half of the tick mark, representing the number Zero (o), was on the negative side and the other half of was on the positive side. Realizing this method of using the whole tick mark, that represents the number Zero (o), is not the correct way of taking an accurate measurement.

However, by dichotomizing or splitting only the tick mark that represents the number Zero (o) on a number line, into two equal tick marks made a negative number Zero (o) and a positive number Zero (o). This Law of Measurement of mine is the correct method of taking a more accurate measurement.

My Law of Measurement applies to the United States customary units, the International System of Units and with the imperial units or the imperial system for taking a more accurate measurement.

Keeping it simple, as examples, there are 8 figures below. Each figure has two illustrations measuring on a number line using Real Numbers.

The top illustration using the stand method of today for measuring involves the use of the whole tick mark, colored in half black and half red, produces an approximation symbol.

The bottom illustration, using my Law of Measurement, splits the tick marks into two equal halves. The negative half side of the tick marks is colored in red and the positive half side of the tick marks are colored in black produces an equal symbol.

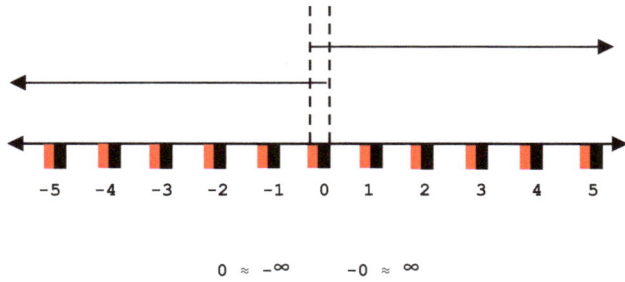

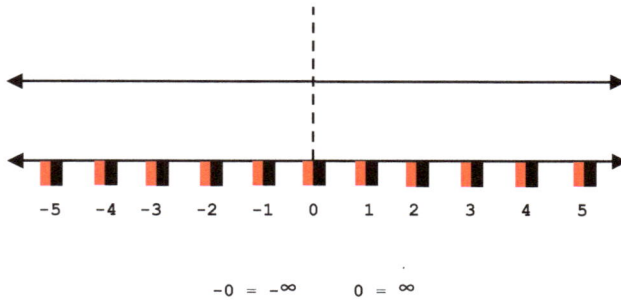

Figure 1

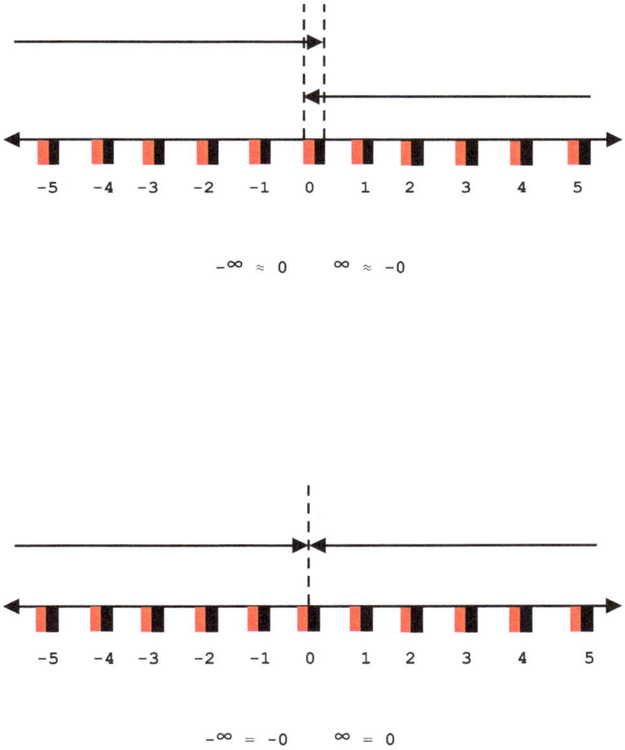

Figure 2

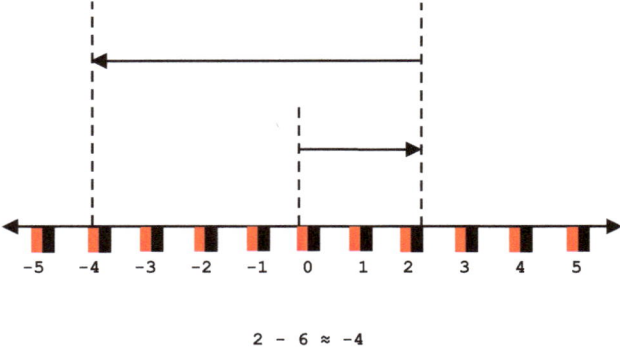

$$2 - 6 \approx -4$$

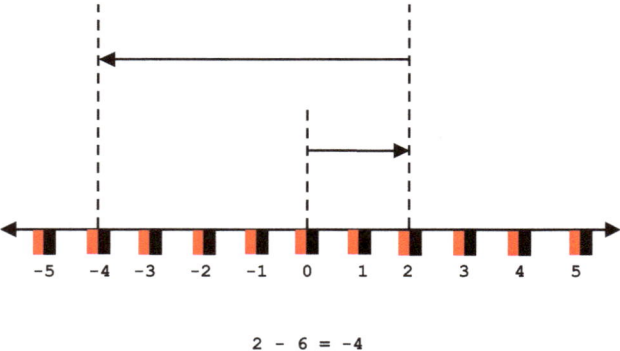

$$2 - 6 = -4$$

Figure 3

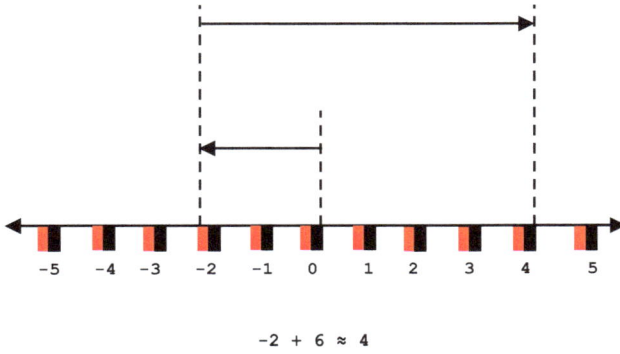

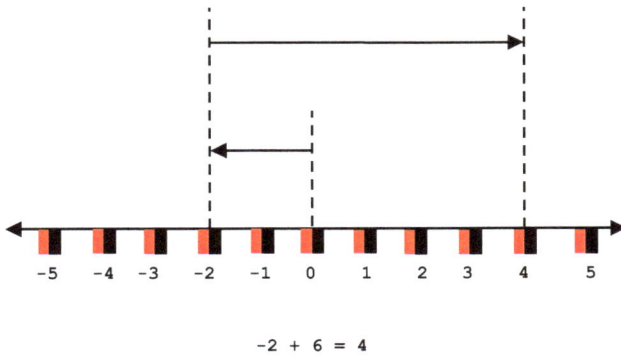

-2 + 6 ≈ 4

-2 + 6 = 4

Figure 4

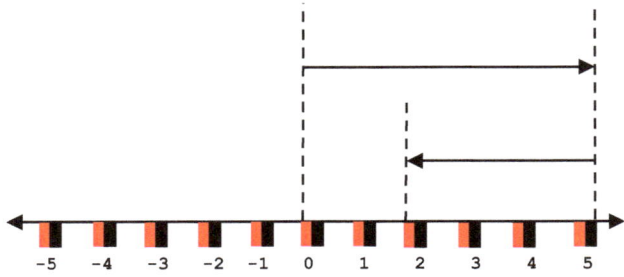

$$5 - 3 \approx 2$$

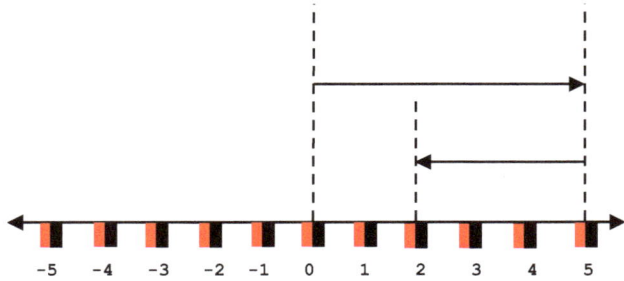

$$5 - 3 = 2$$

Figure 5

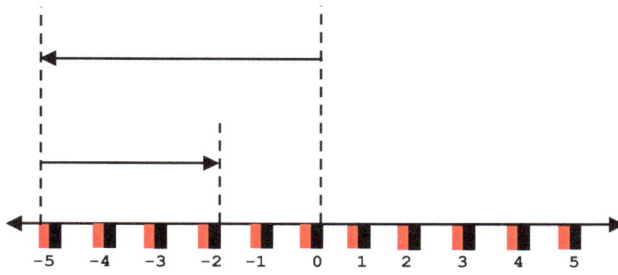

$$-5 + 3 \approx -2$$

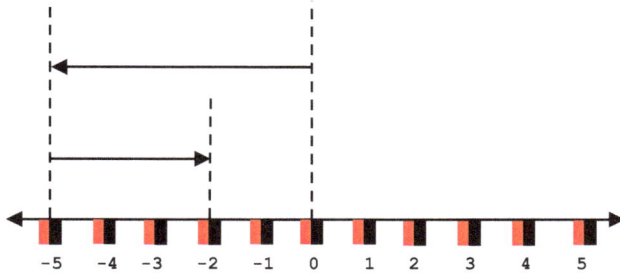

$$-5 + 3 = -2$$

Figure 6

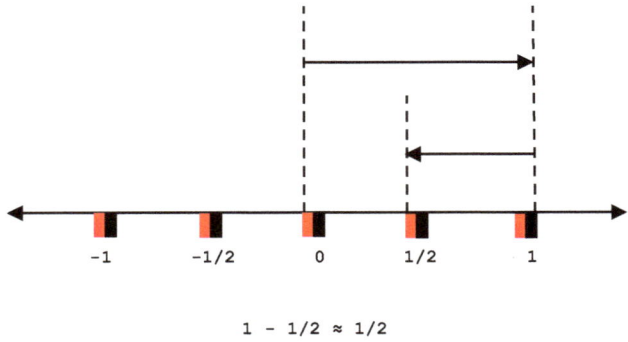

$$1 - 1/2 \approx 1/2$$

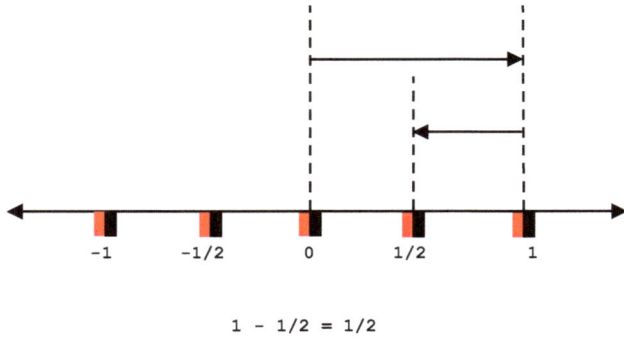

$$1 - 1/2 = 1/2$$

Figure 7

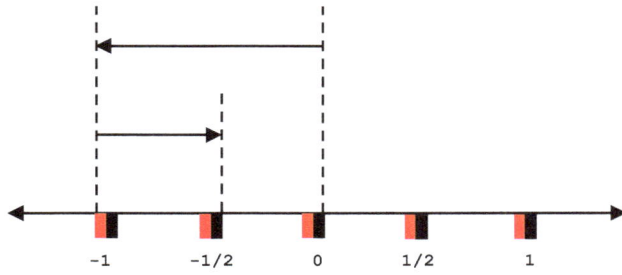

$$-1 + 1/2 \approx -1/2$$

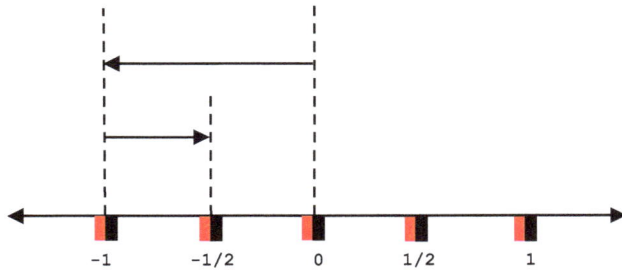

$$-1 + 1/2 = -1/2$$

Figure 8

Chapter 4

Law of Zero

Law of Zero states: Zero (0), is the point of origin of infinity.

Using Real Numbers on a number line, I discovered that the number Zero (0) equals to negative and positive infinity. It is the point of origin in which we can go from the positive side or negative side to infinity and back from infinity to the point of origin, Zero (0).

The number Zero (0) is the only number

that gives both the negative side and positive side a beginning to an infinite number line. No other real number on a number line can do this.

Below, are two illustrations that are drawn to use as examples using Real Numbers on a number line with my Law of Measurement showing how the number Zero (0) is equal to negative and positive infinity. It will also show that the number Zero (0) gives an infinite number line a point of origin to its infinity.

In Figure 1, illustrates on a number line the negative number Zero (0) is the point of origin in which we can go to negative infinity. It also illustrates the positive number Zero (0) is the point of origin in which we can go to positive infinity.

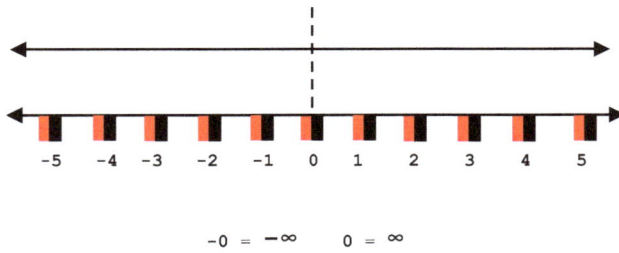

$$-0 = -\infty \qquad 0 = \infty$$

Figure 1

In Figure 2, illustrates on a number line the negative side of infinity coming back to its point of origin, the number Zero (0). It also illustrates the positive side of infinity coming back to its point of origin, the positive number Zero (0).

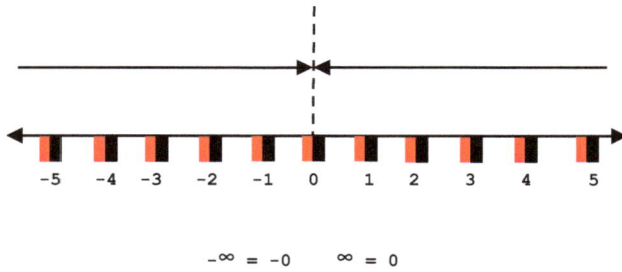

$$-\infty = -0 \qquad \infty = 0$$

Figure 2

Chapter 5

Law of Pi

Law of Pi states: 3.14375 is the true value of Pi.

While conducting my research on the history of Pi, I came across some historical information about the equation circumference divided by the diameter (c/d) has been around for over 4 thousand years and realized that the equation (c/d) was never really utilized in finding the true value of Pi. I asked myself why.

Further research on the history of Pi also showed that the cubit and geometry were the tools used for measurements at that time.

Furthermore, the approximation or false value of 3.14159265... that represents Pi did not derive from the equation (c/d). This false value derived from the use of an infinite amount of regular polygons or the method of exhaustion to measure a perfect round circle in hopes of finding the true value of Pi.

The method of using an infinite amount of regular polygons for finding Pi may be dense enough to look like a circle. However, it is not and because this method is infinite, makes it mathematically impossible to measure.

Thus, makes this method of using an infinite amount of regular polygons for finding the true value of Pi futile. Meaning that for thousands of years, we have overlooked the error of using this method in finding the true value of Pi.

The only standard means of measurement was and is today the use of an infinite amount of regular polygons in finding the true value of Pi and the rest is history. My guess would be that for over four thousand years, this method of using a regular polygon with an infinite number of sides has always been the "norm" in finding the true value of Pi and yet no one has stopped to question it.

That's when I realized, that without a standard linear measurement instrument to measure with, we weren't able to measure the perimeter or the circumference of a perfect circle and why the equation c/d was never really utilized in finding the true value of Pi.

Further researching, on the history of Pi, helped find the vital information that led to the discovery of the Law of Pi. It also gave the understanding, in the origin of the false value that represented Pi.

I was able to ground myself, to see where I needed to go, to help further my research in correcting this error by finding the true value of Pi.

I was able to gain the knowledge and wisdom of understanding why the equation c/d will give the correct or true value of Pi and why the method of an infinite amount of regular polygons will always give an approximation or a false value of Pi.

For the sake of argument, I will be using the phrase, "perfect round circle", to represent the word, "circle" so as not to be confused with any other circles that may appear to look like a circle but are not.

In this experiment, we will apply the Law of Measurement, on a standard linear yardstick.

I proceeded with the experiment by using a straight 47-inch standard linear instrument or yardstick made of Plexiglas. For the 10-inch diameter, I drew a straight line, and measured it from the positive half-side of the inch tick mark, that represents the Zero (0) inch, to the negative half-side of the inch tick mark, representing the positive side of the 10-inch.

I also made sure that the thickness of the flat tip lead, used, for drawing a diameter, was the same thickness as the dichotomized inch tick mark, which is used with Real Numbers, on a standard linear yardstick.

Working with the beam compass, I used the flat tip lead for drawing a perfect circle, in this experiment, rather than using an ink tip pen to avoid compromising the smooth, and unblemished surface.

I carefully placed the flat needle tip end of the beam compass, between the middle negative and positive inch tick mark, that represented the number 5-inch, and placed the flat tip lead end on the positive half-side of the inch tick mark, that represented the Zero (0) inch.

Drawing a perfect circle, using the beam compass, I made sure to start with the outer end of the flat lead tip matched the positive half-side of the inch tick mark represents the Zero (0) inch and matching the negative half-side of the inch tick mark representing the 10-inch.

This method helped to make sure that the circumference of a perfect circle matched both ends of the diameter, and using a standard linear instrument to take correct measurements on the circumference of its perfectly round smooth surface of the circle.

I used a specially modified standard linear instrument designed in making it easier for me to physically take correct measurements of the circumference or around the perimeter of a perfect round circle, and it measured exactly 31 14/32-inches in circumference. I then divided the 31.4375-inch circumference by its 10-inch diameter and came to exactly 3.14375 for the true value of Pi.

This primitive low-tech experimental process of drawing a perfect circle and taking the correct measurements of the circumference was much less expensive. However, there may be other more advanced high-tech means of measuring the circumference of a perfect circle in finding the true value of Pi.

The little experiment confirmed that 3.14375 to be constant and finite. In conclusion, 3.14375 is the true value of Pi. Eureka!

Below are 9 Figure with illustrations using the Law of Measurement and Real Numbers on a number line in finding the true value of Pi.

Figure 1

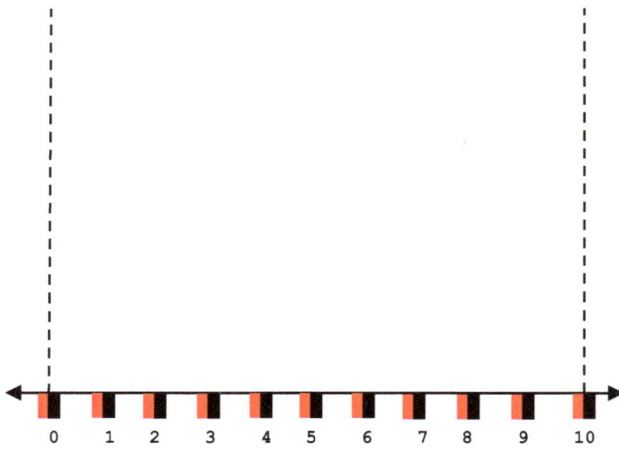

Figure 2

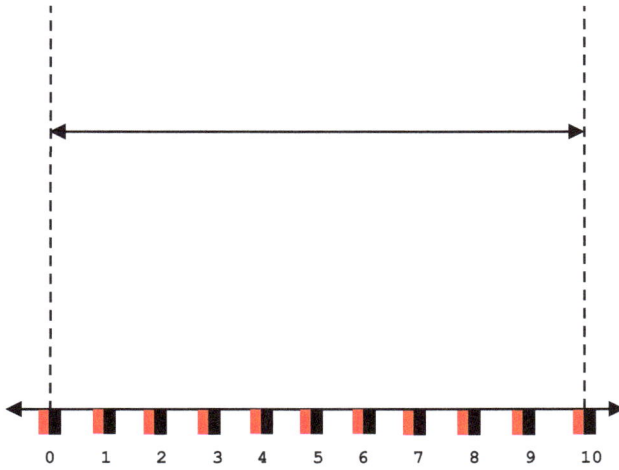

Figure 3

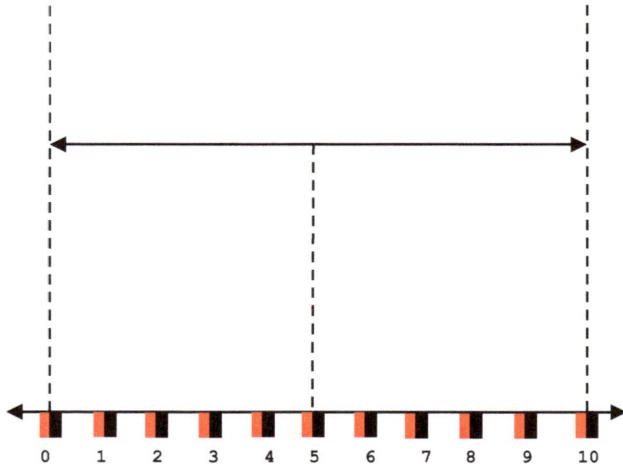

Figure 4

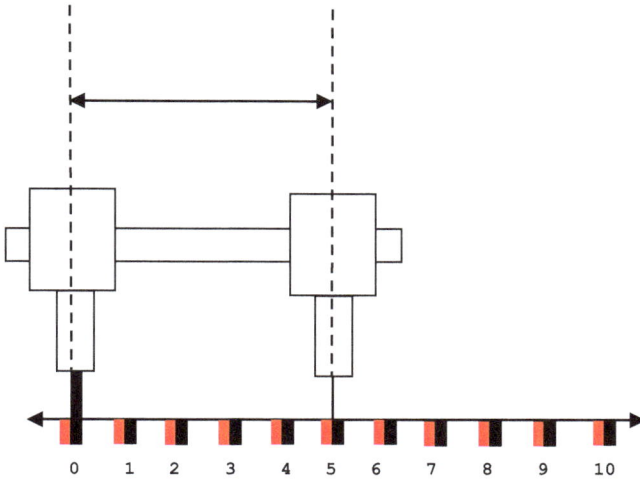

Figure 5

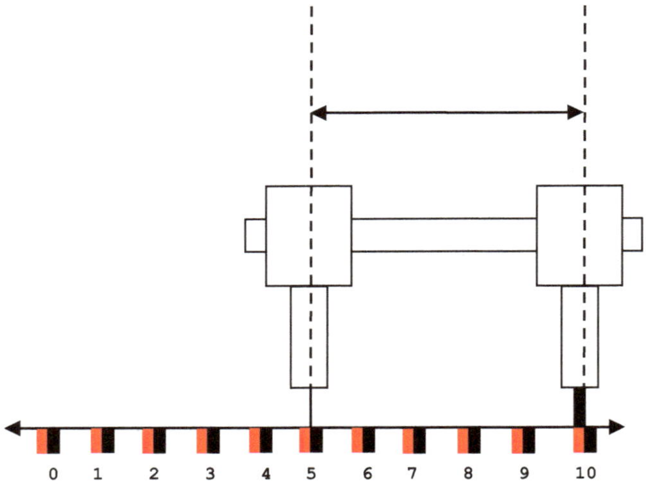

Figure 6

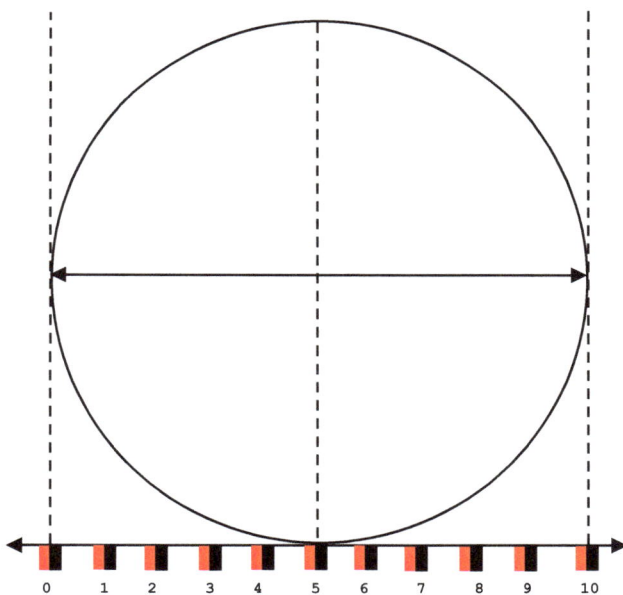

Figure 7

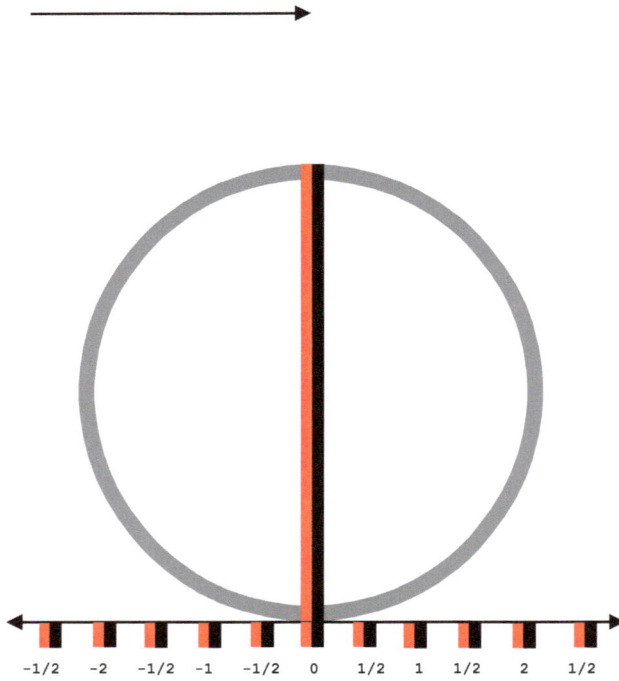

Figure 8

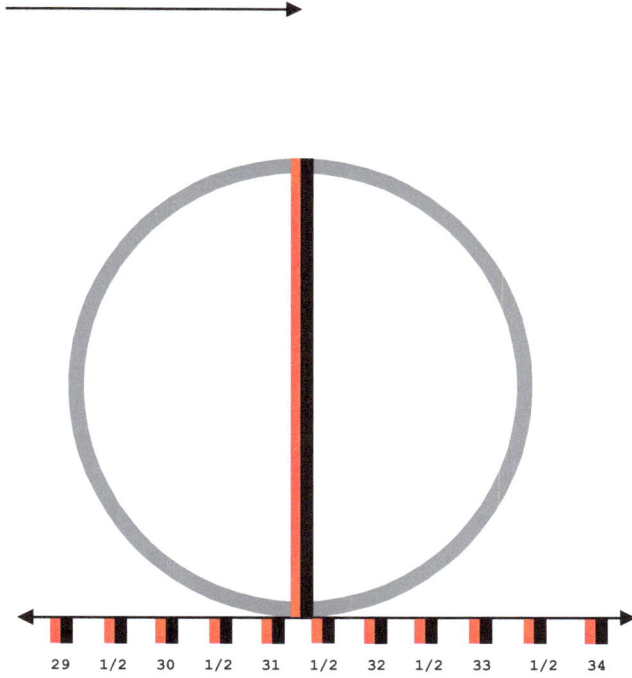

Figure 9

Chapter 6

Law of Kinetic Energy

Law of Kinetic Energy states: An animated body that puts itself into motion or at rest, by applying its own internal energy, remains constant as kinetic energy.

The Law of Kinetic Energy is about all living things that have the ability or the free will within itself to initiate self-motion or come to rest.

The point I'm making here with this new law of physics is not whether or not the energy is work, thermal, or transfer related, but rather HOW a

body was able to be put into motion. The question

being was the energy source an external or internal

that put the body into motion or rest. This is what

differentiates, whether it is kinetic energy or

potential energy.

Chapter 7

Law of Potential Energy

Law of Potential Energy states: An inanimate body at rest or put into motion, by an external source of energy remains constant as potential energy.

The Law of Potential Energy is about all none living things that do not have the ability or the free will within itself, like earth, air, water, or even fire, to initiate the self-motion or come to rest.

The question being was it the applied energy

external or internal that put the body into motion

or rest. This is what differentiates, whether it is

kinetic energy or potential energy.

Chapter 8

Law of Creation

Law of Creation states: "Nothing created can be greater than its creator."

That everything whether born, created, manufactured or came into existents remains constant in every way just as they were designed to be. Meaning, having the ability to be only what it or they were meant to be when it or they were born, created, manufactured or came into existents and nothing more.

44

My research on creation has led me to discover the Law of Creation, which explains that all things created, by design, will stay and forever remain constant in its natural order.

For example, All living things may have the ability to learn, understand, feel, adapted, etc, as they were design to do and nothing more. All non-living things have the ability to do nothing.

Living or non-living things cannot be more or less than what they were created to be by their or its creator.

Researching on the understanding of creation lead me to discover the Law of Creation.

Throughout history, we have asked ourselves, what came first and what came after. Was it creation or evolution? The answer to that question is creation. Creation came first and evolution came after. Without creation, evolution cannot exist. Creation is proof that evolution exists. Everything has a point of origin and creation is the origin of evolution and not the other way around.

Chapter 9

Law of Evolution

Law of Evolution states: "Nothing created can naturally evolve from within."

All things created or manufactured by the creator to have the ability to evolve from within will. All things created or manufactured from the original design, not to evolve, by the creator will never have the ability to evolve from within.

While doing scientific research on evolution lead to the discovery of the Law of Evolution, which explains that all things created can only evolve by an outside modification process. Thus, allowing something to evolve into something else. Without this outside modification process, all things created, in their original design, will forever remain constant within their natural order meaning un-evolved.

Throughout history, we have asked ourselves, what came first and what came after. Was it creation or evolution? The answer to that question is creation. Creation came first and evolution came after. Without creation, evolution cannot exist. Evolution is proof that creation exists. Everything has a point of origin and creation is the origin of evolution and not the other way around.

In conclusion, Evolution itself was created to alter creation.

Chapter 10

Law of Energy

I, like many others in the past, had asked, "What is energy"? No one really knew what energy was until now. I discovered that particle-less energy is the only thing that consists of all of the ingredients in creating a mass or massless particle.

My research leads me to the discovery of what is Energy. It leads me to understand and learn what Energy really is. I call my discovery, "Law of Energy".

Law of Energy states: Energy is the creator that creates mass and massless particles in our universe.

This discovery also lead me to understand and learn that Energy is the common denominator linking all things like animate and inanimate objects, mass, particles, acceleration, motion, the Fundamental Forces, Gravity, Magnetic, Magnetic Field, Electro Magnetic Field, Light, etc., for what it, it would become inert or nonexistent.

Energy is the common denominator that links anything and everything in our universe.

Energy can exist with or without a particle. This means that energy, without a particle, has no shape, form, or structure. Removing the energy from a mass or massless particle will cause it to no longer exist.

Energy is calculated to show all existing entropy. Space, matter and time are all products of an initially applied energy.

Creating a photon particle is as easy as striking two flints together to cause a spark. Because photons are created when energy is applied to created heat.

I also discovered that when striking two objects together, light or a photon particle is created. Striking or smashing two objects or particles together will create other mass or massless particles.

Energy is not to be confused with Kinetic or Potential Energy. Energy, in its original state, itself is particle-less whereas Kinetic and Potential Energy are not.

Energy is the foundation in creating particles. So, what is energy? Scientifically speaking, Energy is the creator in creating all particles.

Chapter 11

Symbol for Massless Particles

Base on my scientific research for a symbol to represent massless particles, I discovered and created a new symbol representing the massless particles that have little to no mass. My new symbol for massless particles is denoted by $\bar{m}$ ("m-bar") as illustrated on figure 1.

m

Figure 1

www.ingramcontent.com/pod-product-compliance
Lightning Source LLC
Chambersburg PA
CBHW041715200326
41519CB00001B/167